RAIL 17 PORTFOLIOS

BR in the Midlands

Compiled by Andrew Fell

First published 1991

ISBN 0 7110 1991 6

Published by

IAN ALLAN LTD

Terminal House Shepperton TW17 8AS
Telephone: Walton-on-Thames (0932) 228950
Fax: 0932 232366 Telex: 929806 IALLAN G
Registered Office: Terminal House Shepperton TW17 8AS
Printed by Ian Allan Printing Ltd at its works at Coombelands in Runnymede, England

Front cover:
No 86229 *Sir John Betjeman* speeds north on 24 January 1990 near Shilton between Rugby and Nuneaton. The Oxford Canal can be seen to the fore.
Mamiya 645J 80mm Sekor C Kodachrome 64 1/250, f5.6

Right:
The bridge at Cossington has been used for countless railway photographs. By 1987 about the best you could hope for at this location was a pair of Class 45s such as No 45150 and No 45052 seen here on the afternoon Derby-St Pancras parcels. Although frequently double-headed, a pair of 'Peaks' on this service was unusual. 24 September 1987.
Pentax Spotmatic 50mm Takumar Kodachrome 64 1/1000, f2.8

Introduction

Although there is no official definition of the area that could be called the Midlands, all of the photographs in this book were taken within about a 60-mile radius of Birmingham. The variety of traffic within the area is great, with the West Coast main line forming the trunk route in the region and the Midland main line to Leicester, Nottingham and Derby also of importance. The East Coast main line has also been included although this is very much at the periphery. The Western Region influences the Midlands to the southwest of Birmingham, both in terms of motive power and lineside features. Another line worthy of mention is the North East/South West route which is particularly busy, especially north of Birmingham.

Selecting photographs for this book was a difficult task as I wanted to include the greatest variety in terms of locomotive classes, traffic and different lines, whilst, at the same time, producing a selection of photographs that was interesting, scenic and had an emphasis on the more exciting workings in the region. I make no apologies for the under-representation of Sprinters, DMUs, and HSTs, or for the large number of freight trains and particular classes such as the Class 20s.

Like other areas of Britain's railways, the Midlands has seen numerous changes in recent years with several locomotive classes disappearing, but new locomotives coming into service and some freight lines reopening for passenger traffic. The future is certainly looking good.

I would like to thank all those whose co-operation helped me to take some of the photographs in this book: various British Rail employees, the power station managers and the other organisations who spared me the time when they were under no obligation to do so.

Andrew Fell July 1990

Left:
With the town in the background, Nos 20160/056 reverse their MGR train past the coalstock and into Rugeley power station. The train has come down from either Trentham or Calder Valley colliery, both near Stoke-on-Trent. Having diverged from the WCML on to the Cannock line, the train has to reverse to gain access to the power station. After unloading, the engines will run round their train, reverse on to the Cannock line and proceed out on to the WCML again. 8 May 1989.
Mamiya 645J 80mm Sekor C
Kodachrome 64 1/250, f5.6

Right:
Sundays are usually a low point in the railway week but in 1987 they provided an excellent opportunity to capture a Class 45 in action on the Midland main line. The train is the 1C22, 12.15 SuO Derby-St Pancras. Although diagrammed for a Class 47, one rarely appeared. No 45121 works the train past Newton Harcourt just south of Leicester on 16 August 1987. The locomotive carries the unofficial name *Pegasus* painted on the side.
Pentax Spotmatic 50mm Takumar
Kodachrome 64 1/1000, f2.8

Above:
No 56110 hauls empty Redland stone hoppers back to Mountsorrel past Kibworth Harcourt. Like several freight services, this train, when loaded, works south along the freight only line through Corby in order to avoid the steep gradients on this route. 5 April 1990.
Pentax Spotmatic 135mm Carl Zeiss Kodachrome 64 1/250, f8-5.6

Above:
At the time of this photograph, 3 September 1986, Harringworth signalbox was open, but only until 16.00. After an uneventful afternoon, the signalbox was closed just before three southbound workings passed in succession, operating over one extended section from Manton Junction to Corby. The sun came out for the third of these trains, No 31260 on a ballast working from Loughborough.

Pentax Spotmatic 50mm Takumar Kodachrome 25 1/500, f2.8

Above:
The freight only line north of Corby offers some magnificent locations. Nos 37668/667, seen here, have just emerged from the mile long Corby Tunnel and journey north with the morning Corby-Lackenby 'tubeliner' steel empties. The route to Manton Junction includes a series of 'S' bends, tunnels and the impressive Welland Viaduct. 18 August 1989.
Mamiya 645J 80mm Sekor C Kodachrome 64 1/500, f5.6-4

Right:
Vic Berry's scrapyard in Leicester has disposed of numerous locomotives in recent years. A few years ago, the future of BR's only 100mph diesel locomotive class looked quite secure but views such as this provide a prompt reminder of how quickly the railways and BR's policies are changing. The locomotives in question are Nos 50047/022. Also present are Nos 47103 and 25202. 3 May 1989.
Mamiya 645J 80mm Sekor C Kodachrome 64 1/125, f8

47 103

Above:
A delightful triple header passes Cossington in the late afternoon of 11 July 1989. No 37274 leads Nos 20171/195 on a rather tatty mixed rake of stone hoppers.
Mamiya 645J 80mm Sekor C Kodachrome 64 1/500, f4

Right:
No 56110 is about to pass under the A6006 roadbridge near Normanton on Soar with the 7085, 13.23 Toton-New Bank-Northfleet MGR service on 3 August 1987. The Class 56 or 58 diagrammed for this working continued right through to the Southern Region, thus providing the unusual sight of a Midland Region Type 5 locomotive on Southern metals.
Mamiya 645J 80mm Sekor C Ektachrome 64 1/500, f4

56110
110

Left:
Nos 20085/059 wait to unload their train at Ratcliffe power station. Both tracks through the unloading stage are in operation as can be seen by the MGR wagons on the adjacent line. Ratcliffe is one of the largest power stations in the country and the coal traffic is almost continuous through the day. 28 July 1989.
Mamiya 1000S 80mm Sekor C
Fujichrome 100 1/30, f16

Right:
On the same day as the last photograph, No 58034 *Bassetlaw* progresses slowly through the unloading stage at Ratcliffe. The power station complex can be seen to the left. On unloading, the train will continue round a large loop before leaving the power station at the same junction it entered from, just to the south of Redhill Tunnel. (Photographs courtesy of PowerGen plc, Ratcliffe-on-Soar power station.)
Mamiya 645J 80mm Sekor C
Kodachrome 64 1/250, f5.6-4

Above:
Class 20s Nos 20142/196 reverse their train under the loading stage at Calverton colliery in the Nottinghamshire coalfield. The train is the 6A32 to Ratcliffe which runs on alternate weekdays. The same train will return in the afternoon to form a second working to the same destination. 30 March 1990.
Mamiya 645J 80mm Sekor C Kodachrome 64 1/125, f8

Right:
A simple but pleasant location is enhanced by a patch of cloud in an otherwise clear sky. Nos 20054/124 leave Bottesford Junction near Grantham and head for the coast with the 07.18 SO Derby-Skegness. Class 20s made a welcome return to this route in 1988 and two pairs could be seen on a good Saturday. 3 September 1988.
Mamiya 645J 80mm Sekor C Kodachrome 64 1/500, f4

Left:
In complete contrast to the previous photograph, No 91005 glides southwards at the rear of the 17.10 Leeds-Kings Cross service on a glorious sunny evening. The location is Stoke Tunnel which lies between Grantham and Peterborough. Fortunately, overhead wires have not spoilt this location too much. 26 May 1990.
Pentax Spotmatic 135mm Carl Zeiss Kodachrome 64 1/250, f5.6

Right:
This location at Stonebroom lies between Clay Cross Junction and Alfreton on a line better known for freight traffic to and from Toton. In a fairly short-lived patch of sunshine, Sprinter No 156466 speeds south with an afternoon Provincial Sector working. 25 September 1989.
Pentax Spotmatic 50mm Takumar Kodachrome 64 1/1000, f4-2.8

56466

Above:
Nos 31238/302 pass under the M1 motorway on the freight only line between Sheetstores and Stenson Junctions. This line sees a lot of traffic to and from Toton and also trains originating from the Humber oil refineries travelling down via Lincoln and Nottingham. This one is probably destined for the terminal at Kingsbury. 17 August 1988.
Mamiya 645J 80mm Sekor C Kodachrome 64 1/500, f5.6-4

Right:
No 58043 heads west at Church Gresley on the Coalville freight line with an MGR train. Mining subsidence is clearly visible in this view. Coal accounts for most of the traffic on this line, destined for Drakelow or Didcot power stations in the main. However, a couple of quarries south of Coalville also generate traffic on revenue earning services and for BR's own departmental requirements. 6 April 1990.
Pentax Spotmatic 135mm Carl Zeiss Kodachrome 64 1/250, f8-5.6

SWAN LIGHT
WON'T TURN A
Dollamore
043

Above:
Elford loop lies between Tamworth and Burton on Trent on the NE/SW main line. In a brief moment of activity, No 37886 on a steel train for the Birmingham area pulls into the loop to allow the passage of an IC125 on the 11.42 Newcastle-Penzance service. The IC125's power cars are Nos 43002/019 and the date is 7 June 1990.
Pentax Spotmatic 135mm Carl Zeiss Kodachrome 64 1/250, f8-5.6

Right:
A summer Saturday on the NE/SW main line at Wadborough between Cheltenham and Birmingham, and the Class 47s and 31s on passenger workings are supplemented by No 37197 on the Tunstead-Margam lime train. 24 June 1989.
Mamiya 1000S 80mm Sekor C Fujichrome 100 1/1000, f4

088

Left:
A welcome spell of cold weather in February 1991 created several days of ideal sunshine and snow photographic conditions. On 13 February, No 56088 was photographed hauling a southbound MGR train approaching Wichnor Junction between Burton on Trent and Tamworth. This is the junction for the now infrequently used freight line to Brownhills.
Pentax Spotmatic 135mm Carl Zeiss
Kodachrome 64 1/500, f5.6/4

Right:
Tints of autumn portray the time of year in this photograph of No 37278 on 24 October 1988. The Cardiff-based engine is hauling steel empties back towards South Wales, passing the unusually named village of Up Hatherley, between Cheltenham and Gloucester. Freight activity on this line can fluctuate, with very busy spells or lengthy periods with no traffic whatsoever.
Mamiya 1000S 80mm Sekor C
Ektachrome 100 1/500, f4

47 573
STANDARD
Network SouthEast

Left:
The semaphore signals at Worcester Shrub Hill witness the departure of No 47573 *The London Standard* on the 13.45 Paddington-Hereford service on Sunday 25 March 1990. By this time BR was beginning to achieve its goal of uniform liveries on both locomotives and rolling stock. The tranquil atmosphere at stations like this is in stark contrast to the likes of Birmingham New Street and the attractive surroundings more than compensate for the fewer number of workings.
Mamiya 645J 80mm Sekor C
Kodachrome 64 1/250, f5.6

Right:
At the beginning of 1990 the Worcester area became a target for the elimination of semaphore signals. The replacement colour light signals are already in position, as can be seen from this view of Norton Junction as IC125 Nos 43169/032 take the spur to join the main line to Cheltenham with the 14.14 service from Great Malvern to Paddington, the 'Cheltenham Spa Express', on 26 March 1990.
Pentax Spotmatic 135mm Carl Zeiss
Kodachrome 64 1/250, f8-5.6

Above:
During the summer of 1989 the Oxford-Worcester line saw diversions every Sunday of trains from Birmingham that were usually routed via Coventry and Banbury. The result was a revival of loco-hauled workings throughout the day. The traction was predictably Class 47s, although a Class 31 was not unknown. The first southbound working of the day, the 07.10 Manchester Piccadilly-Poole service, passes Evesham in the charge of No 47533. 25 June 1989.
Mamiya 645J 80mm Sekor C Kodachrome 64 1/500, f5.6-4

Above:
With less than a year to go in service, celebrity Class 47/4 No 47406 passes Moreton-in-Marsh with the 12.20 Wolverhampton-Reading service, a service which is not scheduled to stop at this station. This was the only original Class 47/4 to make the transition into InterCity livery. The semaphores at Moreton-in-Marsh mark the beginning of a short double track section on this route which is otherwise essentially single track all the way from Hereford to Oxford. 24 September 1989.
Mamiya 1000S 80mm Sekor C Ektachrome 100 1/500, f5.6

Oxford
L201

Left:

Network SouthEast liveried DMU No L201 heads south away from Aynho Junction and past Souldern Viaduct. The train is the punctual 14.45 Banbury-Oxford service and the date is Saturday 23 September 1989.

Pentax Spotmatic 50mm Takumar
Kodachrome 64 1/1000, f2.8

Right:

Signs of spring blossom have already arrived by the date of this photograph, 17 March 1990. No 47556 is hauling the slightly late running 07.45 Newcastle-Poole service, 'The Northumbrian'. The location is Old Milverton on the single track Coventry to Leamington Spa route. This line has remained single since reopening to passenger services despite the high levels of traffic.

Mamiya 100S 80mm Sekor C
Ektachrome 100 1/500, f5.6

Right:
A very smart Class 47 and matching rake of coaches present a positive image for InterCity. The locomotive is Class 47 No 47821 and the train is the 07.18 Manchester Piccadilly-Brighton service on the May Day bank holiday, 7 May 1990. By this time Class 47s had a virtual monopoly on passenger services over this route and a week after this photograph was taken their domination was to be complete as the last Class 50 working disappeared with the introduction of the new timetable.
Pentax Spotmatic 135mm Carl Zeiss Kodachrome 64 1/250, f8-5.6

Far right:
Fenny Compton is a delightful location that boasts upper and lower quadrant signals side by side. The location lies within the London Midland Region but this is an inter-regional route and No 50027 *Lion* adds to the Western Region flavour. The train is the 13.10 Birmingham New Street-Paddington service on 20 August 1987. Direct trains to Paddington have declined in number over this route since the early 1980s.
Mamiya 645J 80mm Sekor C Ektachrome 64 1/500, f5.6-4

50027

Left:
Sunday engineering work sees the 14.22 Wolverhampton-Euston service diverted from its normal route via Coventry and running to Nuneaton under diesel power. This was the last southbound diversion of the day, but the first to produce something other than a Class 47. No 58005 makes a glorious sight as its heat haze distorts Daw Mill colliery between Birmingham and Nuneaton. The use of Driving Van Trailers was standard on Euston-Wolverhampton services by this date, the DVT leading on the southbound working. No 87021 can just be seen bringing up the rear. Note the bullhead rail still in use on this main line. 8 April 1990.
Pentax Spotmatic 135mm Carl Zeiss Kodachrome 64 1/250, f5.6

Above:
On a day marked by sunshine and showers, No 37229 *The Cardiff Rod Mill* draws into Nuneaton on a train of EMU stock, probably originating from Wolverton works. The South Wales engine is well off the beaten track here at a time when Class 37s were extremely rare on this part of the WCML. Today they are regular performers through Nuneaton on stone trains from Peak Forest, near Buxton, to Bletchley. 28 May 1986.
Pentax Spotmatic 50mm Takumar Kodachrome 25 1/250, f4

Above:
This location between Nuneaton and Hinckley is as good a place as any to photograph one of the few freight workings on the Leicester-Nuneaton line. The train is the Three Spires Junction-Toton New Bank MGR empties, headed by Nos 20210/151 on 31 May 1989. Class 58s are also regular performers on this turn though not as exciting as the vintage locomotive class seen here.
Mamiya 645J 80mm Sekor C Kodachrome 64 1/500, f5.6-4

Right:
Sprinter No 156409 passes Croft sidings, between Leicester and Nuneaton, with the 13.25 Birmingham New Street-Harwich Parkestone Quay service on 21 May 1990. Class 56 No 56103 is stationary, waiting to reverse its train of empties into the sidings. The quarry at Croft generates ballast for BR's own use plus granite for English China Clay.
Pentax Spotmatic 135mm Carl Zeiss Kodachrome 64 1/250, f8-5.6

CROFT SIDINGS
156409

Railfreight

Left:
Stratford allocated Class 31s Nos 31190/191 return to home territory as they curve away from Manton Junction and head towards the regional boundary and Stamford beyond. Both locomotives belong to the Departmental Sector and thus are at home on a ballast train such as this. 12 July 1990.
Mamiya 1000S 80mm Sekor C Ektachrome 100 1/500, f5.6

Above:
The Birmingham-Norwich route suffered a great loss in 1988 when the principal passenger services went over to Sprinter operation. Such changes may be seen as progress by BR but few enthusiasts have welcomed the invasion of this new generation of rolling stock. No 31449 is seen here departing from Melton Mowbray on 14 September 1987 with the 14.20 Birmingham New Street-Ipswich service. The Birmingham-Norwich line is a classic cross-country line, bisecting a large number of principal routes to London. Each section on the line offers a different selection of freight, for whilst most passenger services travel the length of the route from Birmingham as far as Ely, freight trains are much more localised.
Mamiya 645J 80mm Sekor C Ektachrome 64 1/500, f4

Left:
No 37699 leaves Claydon L&NE Junction having just run round the 6M13 Aylesbury-Washwood Heath Speedlink coal service. On this occasion the train was running three hours early, seen here at 15.43. Minor freight lines, such as Bletchley to Oxford and Aylesbury, are well worth photographing although traffic is rather scarce. The manual signalbox and semaphores had only recently been replaced at this junction with colour lights although a new box was built and is manned. 4 August 1987.
Mamiya 645J 80mm Sekor C Ektachrome 64 1/500, f4

Above:
A pleasant Sunday afternoon sees No 87025 *County of Cheshire* racing towards London with an up express at Soulbury, near Leighton Buzzard, on 12 February 1989.
Pentax Spotmatic 50mm Takumar Kodachrome 64 1/1000, f2.8

WHILTON MARINA
HIRE BOATS
PLEASE TIE
UP OUTSIDE
MARINA

Left:
The sun sets over the Grand Union Canal at Whilton Marina, between Kilsby and Weedon, at precisely 15.23 on 23 November 1989. Even as late as this in the year there is still a fair amount of boating activity on the canal. The unidentified Class 90 is hauling a short southbound train of 'Ford' vans.

Pentax Spotmatic 50mm Takumar Kodachrome 64 1/250, f2.8

Below:
The sweeping curve to the north of Weedon provides the setting for No 86411 on a London bound express. The Grand Union Canal (in the background) passes the WCML in a number of places before heading north to Leicester, Loughborough and then on past Ratcliffe and Toton. 31 October 1988.
Mamiya 1000S 80mm Sekor C Ektachrome 100 1/500, f5.6-4

Above:
The race is on as No 90012 heads north alongside the M1 in Northamptonshire, close to Watford Gap service station. DVT operation was well established by this time and allows for an entire train formation of Mk 3 stock, not previously possible when brake coaches were needed. The photograph is dated 3 November 1990 at precisely 8.13am.
Pentax Spotmatic 135mm Carl Zeiss Kodachrome 64 1/250, f5.6/4

Right:
Two short tunnels are visible in this view of No 31403 heading a short southbound Departmental Sector train on the Northampton loop, near Watford village. Class 31s became commonplace on the southern end of the WCML following the demise of the Class 25s. Since the Class 31/4s lost the majority of their passenger diagrams they have become the norm in the area, although they seldom appear on revenue earning freight. 31 May 1990.
Pentax Spotmatic 135mm Carl Zeiss Kodachrome 64 1/250, f8-5.6

Above
Diverted from its usual route, No 85025 takes the Northampton loop line and is seen here at Long Buckby with a northbound passenger train. Diversions over this line are commonplace at weekends and even on weekdays. The distinctive yellow of oil seed rape fields can be a little overpowering but by the date of this photograph, 27 May 1989, the colour in the fields had toned down. At this time the Class 85s could still be seen on a fair number of passenger diagrams but their Class 81 sisters were extremely rare performers.
Mamiya 1000S 80mm Sekor C Ektachrome 100+ 1/1000, f4

Right:
No 47555 speeds through Tile Hill on 9 April 1990 with the 08.18 Manchester Piccadilly-Poole service. This line is particularly busy today with half-hourly Wolverhampton to Euston expresses combining with hourly Birmingham to Euston slow trains, locals to Coventry and hourly inter-regional trains such as this one. Freight traffic has, however, become increasingly infrequent.
Pentax Spotmatic 135mm Carl Zeiss Kodachrome 64 1/250, f8-5.6

CY
60
InterCity

Below:
No 50024 *Vanguard* storms through Berkswell on 9 April 1990 with the 11.18 York-Paddington service, which it has worked from Birmingham. At this time this was the only Class 50 passenger train to operate under the wires, as the locomotive only worked as far as Coventry on its northbound diagram, the 09.15 Paddington-Manchester service, and then proceeded light engine for its afternoon return duty.
Pentax Spotmatic 135mm Carl Zeiss Kodachrome 64 1/250, f8-5.6

Right:
Construction cranes dominate the skyline of central Birmingham in this view of DMU No T035 at Birmingham Moor Street station. Until recently the terminus for these trains, Moor Street is now a rationalized through station for trains originating at the reopened Snow Hill. The train here is the 14.30 Snow Hill to Stratford-upon-Avon service on 22 August 1990.
Mamiya 1000S 80mm Sekor C Ektachrome 100 1/125, f11-8

ARKS & SPENCER
1
Mocr Street
T035

Right:
Lichfield City station offers a pleasant combination of semaphore signals and traditional station architecture, both of which are becoming less common in the West Midlands area. DMU No T410 heads away from the camera with the 16.02 Lichfield Trent Valley-Longbridge service on 22 August 1990. This route is to be electrified through to Redditch.
Mamiya 1000S 80mm Sekor C
Ektachrome 100 1/125, f11-8

Far right:
EMU No 310102 enters Wolverhampton in preparation to form a stopping service to Walsall. Trains between Wolverhampton and Birmingham are abundant with express services to Euston every half-hour in addition to other main line trains, local workings and a good few freight trains. 22 March 1990.
Mamiya 1000S 80mm Sekor C
Ektachrome 100 1/250, f8-5.6

310
102

023

Left:
The M6 motorway and gas storage towers set the scene in the heart of industrial Birmingham. The location is Washwood Heath and the train is yet another MGR working in the hands of a Class 58. No 58023 is caught in low winter sunshine on 18 January 1990 shortly after the morning frost had melted.
Pentax Spotmatic 50mm Takumar Kodachrome 64
1/500, f4-2.8

Right:
One of the busier freight lines in the region is that from Walsall to Water Orton. Bescot is probably the destination of this westbound Departmental Sector train as it nears Walsall behind a very smart No 31460. The trees are just beginning to come into leaf by this date, 4 May 1989.
Pentax Spotmatic 50mm Takumar Kodachrome 64
1/500, f4

Above:
This interesting location is Haughton between Shrewsbury and Gobowen. The signalbox is usually closed now but is still ready for use should it be needed in an emergency. No 47200 passes one of the lower quadrant signals with the Washwood Heath-Gobowen Speedlink coal service on 4 July 1989. The following day the network came to a stop because of one of the one day strikes that occurred on several Wednesdays during that summer.
Mamiya 1000S 80mm Sekor C Fujichrome 100 1/500, f5.6

Right:
During the summer of 1989, Sprinter shortages resulted in Class 37 haulage on the Crewe-Cardiff route, thus providing a reprieve before the impending loss of loco-hauled workings altogether. The Cardiff-based Class 37/4s were used on trains to and from Liverpool with the alternate Manchester trains formed by Sprinters. No 37430 *Cwmbrân* heads the 09.15 Liverpool-Cardiff service past All Stretton. Class 37 haulage survived into 1990 although on a smaller number of diagrams. 3 July 1989.
Mamiya 1000S 80mm Sekor C Fujichrome 100 1/1000, f4

37430

Left:
The view from the walls of Shrewsbury Castle offers a magnificent panorama of the southern approach to the station and part of the town. Some semaphores have disappeared from Shrewsbury in recent years but several fine examples can still be seen in this view as well as the imposing signalbox. Note also the oil tanks stabled at Shrewsbury Abbey sidings. The afternoon Llanwern-Dee Marsh steel train approaches the station behind No 37886 on 22 March 1990. No 37428 is also present, stabled between duties. Since the photograph was taken the sidings at Shrewsbury Abbey have closed.
Mamiya 645J 80mm Sekor C Kodachrome 64 1/250, f5.6

Above:
Cosford is situated between Wolverhampton and Telford and enjoys a reasonable variety of traffic, including the regular coal trains to Ironbridge power station. Semaphore signals survive at several places on this line. No 47500 *Great Western* passes the loop lines at Cosford with the 11.04 Shrewsbury-Euston service on 27 March 1990. Note the inclusion of a Driving Van Trailer in the formation by this date.
Pentax Spotmatic 50mm Takumar Kodachrome 64 1/1000, f2.8

Right:
Class 20s Nos 20172/028 approach the unloading stage at National Power's Ironbridge power station. Standard procedure is for the hoppers to be automatically discharged while the train progresses at a constant 1/2mph. However, on this occasion the hopper doors were being opened manually. This garishly painted pair of Class 20s belonged to the Departmental Sector. 6 November 1989.
Pentax Spotmatic 50mm Takumar Kodachrome 64 1/8, f4

Far right:
When coal trains arrive at the power station in close succession it is often necessary to hold one while the other is unloaded. Such is the case here as Nos 20006/117 pause under the powerful floodlights at Ironbridge power station. The blurred outline of another pair of Class 20s can be seen behind as they make progress forward unloading. Although normally only used for freight services, the Ironbridge line has seen limited use for special passenger trains serving the Ironbridge Gorge Museum at Coalbrookdale over the past few summers. 6 November 1989.
Mamiya 645J 80mm Sekor C Kodachrome 64 40secs, f5.6

20 066

Left:
Thornaby-based Steel sub-Sector locomotives Nos 37507/508 haul the 6M13 Tees Yard to Etruria towards the Potteries at Church Leigh on the Crewe-Derby line. Semaphore signals survive on this route, this one being controlled by Church Leigh signalbox. The box is situated some distance down the line, overlooking a level crossing at the site of the former station. 22 May 1989.
Pentax Spotmatic 35mm Flektogon Kodachrome 64 1/500, f5.6-4

Right:
On 22 May 1989, No 47196 *Haliotidae*, resplendent in the new Railfreight livery complete with Sector markings, works an eastbound oil train at Bramshall near Uttoxeter. Freight traffic on this line can be disappointing, but this exceptionally busy day produced oil, MGR, steel and departmental trains, and even a rake of London Underground stock.
Mamiya 1000S 80mm Sekor C Fujichrome 100 1/500, f5.6

47 196

Above:
The main line between Crewe and Stafford is not known for its panoramic locations, but this spot, near the village of Hill Chorlton, is an exception. No 87018 *Lord Nelson* heads a southbound working on 11 October 1989. Despite the threatening sky, the abundance of trains on this line ensured that some sunlit photographs were taken.
Mamiya 645J 80mm Sekor C Kodachrome 64 1/500, f4

Right:
The delightful setting of Little Haywood, near Stafford, is the location for this view of Class 20s Nos 20071/182. The pair are approaching Colwich Junction with an afternoon MGR train from Trentham to Rugeley power station. On such a short circuit the locomotives will spend considerably more time engaged in loading/unloading than main line travel. This particular stretch of the WCML sees relatively little traffic, being used principally by Manchester-Euston expresses avoiding Crewe, Stafford and Birmingham. 20 June 1989.
Pentax Spotmatic 35mm Flektogon Kodachrome 64 1/500, f5.6

Above:
A view from the roof of Rugeley 'B' power station shows Nos 20182/071 reversing a load of MGRs from Trentham past the unloading stage before pulling forward again to discharge. They will have to wait for Nos 20058/082 to unload first as the leading locomotive creeps forward at the beginning of the unloading process. 16 May 1989.
Mamiya 645J 80mm Sekor C Kodachrome 64 1/250, f8-5.6

Right:
This panoramic view was also taken from the roof of Rugeley 'B' power station. Nos 20159/045 enter Rugeley with an MGR from Essington near Cannock. Nos 20088/187 can be seen behind with a train from Trentham that will reverse into the power station. The WCML can be seen in the background.
Mamiya 645J 80mm Sekor C Kodachrome 64 1/250, f8-5.6

GTB 4

Above:
Class 20s Nos 20135/080 emerge from the unloading stage at Rugeley on a misty May morning with an MGR coal train. At this time the Class 20s dominated workings to this power station, with only the odd Class 58 putting in an appearance. This dominance will soon cease with the introduction of the new Class 60s. 8 May 1989.
Pentax Spotmatic 50mm Takumar Kodachrome 64 1/250, f5.6

Right:
Although not as distinctive as the green applied to some locomotives, the original blue livery on No 86426 makes for a more interesting photograph than the average Class 86. The locomotive was repainted in 1987 in conjunction with the 21st birthday of InterCity. The authenticity of the paint scheme is spoilt only by the large yellow warning panel, which is obligatory today. The locomotive is seen here at Mancetter, near Nuneaton, on an afternoon northbound service.
Pentax 6x7 105mm Takumar Ektachrome 200 1/1000, f5.6

86 426

Above:
In ideal photographic weather, Nos 37515/521 pass Toton with the afternoon Lackenby-Corby 'tubeliner'. This traffic has long been a favourite with photographers, bringing double-headed Class 37s into Leicestershire and Northamptonshire and, these days, interesting combinations of livery are commonplace on the Thornaby-based locomotives. 17 August 1989,
Mamiya 645J 80mm Sekor C Kodachrome 64 1/250, f8-5.6

Back cover:
No 56078 powers south with the 6V76 Cliffe Hill-Hayes Tarmac stone train at Sharnbrook on 24 August 1990. By this time the goods line between Bedford and Glendon Junction had been singled in some places, such as here, or removed altogether, such as north of Wellingborough. This situation could change with the advent of the Channel Tunnel and BR requiring the 'spare capacity' for additional freight traffic. Note the clouds of dust being emitted from the wagons.
Mamiya 1000S 80mm Sekor C Ektachrome 100 1/500, f5.6